Project MIND
<u>M</u>ath <u>I</u>s <u>N</u>ot <u>D</u>ifficult

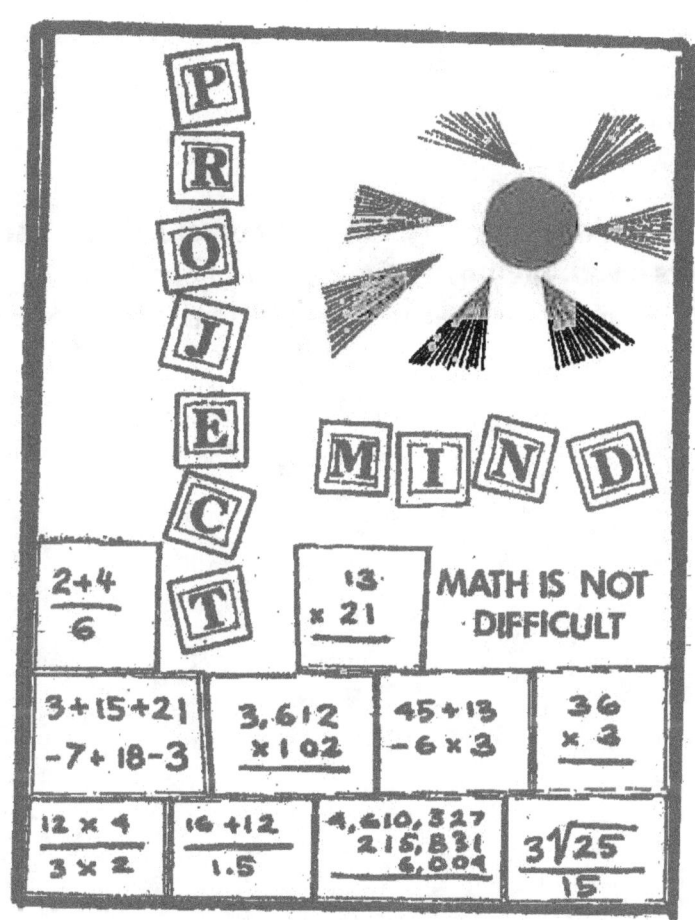

Second Grade
Mental Math Flash Cards
Hui Fang Huang "Angie" Su, Ed.D.
Project MIND, Inc.

The Mental Math Game

The students form two teams and come up to the bells two at a time. Upon looking at the math problem on a yellow card, they solve the problem mentally as fast as they can, usually within three seconds. The winner continues on while the loser moves to the end of his line. To be an intermediate champion, one must respond to three problems in a row correctly. After four intermediate champions are picked (depending on the size of your group, you must make sure that each student had a t least three chances), they are then entered into the second level of competitions with the green cards (more difficult problems.) To be a runner up for the grand champion title, competitors must also respond correctly to three problems in a row. Two runner-ups for the advanced level cards (red) are picked. They now compete for the title. The first person to respond to three red card problems in a row correctly is the grand champion.

Variations:

> The students compete in four areas: Mentally solve math problems with cards (visual aids), mentally solve math problems without cards, word problems, and equations (a string of problems to solve as the reader reads them.)
> The game can be played with your own class, another class, your grade level, or with other grade levels (fourth grade competing against fifth grade, third grade competing against fourth grade, etc.)
> If you have an advanced group, make sure that they use the cards for the next grade.
> Decimals and fractions can be added for third through fifth grade.

Pre-Kindergarten/Kindergarten:

- Level 1 – Yellow Cards: Number identification and shape identification
- Level 2 – Green Cards: Number identification (up to 100), identify the missing number, and adding and subtracting up to 5.
- Level 3 – Red Cards: number sequencing, and adding and subtracting up to 10
- Equations: strings of numbers which add and subtract up to 10
- Word problems: Simple one step, how many items? Adding or subtracting up to 10

First Grade:

- Level 1 – Yellow Cards: Adding and subtracting numbers up to 10
- Level 2 – Green Cards: Adding and subtracting two-digit numbers and adding three digit numbers
- Level 3 – Red Cards: Adding and subtracting three-digit numbers

Second Grade:

- Level 1 – Yellow Cards: Adding and subtracting two-digit numbers

- Level 2 – Green Cards: Adding and subtracting two-digit numbers with carrying and borrowing, and multiplication and division facts
- Level 3 – Red Cards: Adding and subtracting three-digit numbers with carrying and borrowing; two-digit multiplication

Third Grade:

- Level 1 – Yellow Cards: Adding and subtracting two-digit numbers with carrying and borrowing; single digit multiplication and division
- Level 2 – Green Cards: Adding and subtracting three-digit numbers with carrying and borrowing, and two-digit multiplication and division
- Level 3 – Red Cards: Adding and subtracting four-digit numbers with carrying and borrowing; three-digit multiplication and division

Fourth Grade:

- Level 1 – Yellow Cards: Adding, subtracting, multiplying, and dividing fourth grade level problem
- Level 2 – Green Cards: Adding, subtracting, multiplying, and dividing fourth grade level problems that are harder than Level 1
- Level 3 – Red Cards: Adding, subtracting, multiplying, and dividing multi-digit fifth grade level problems

Fifth Grade:

- Level 1 – Yellow Cards: Adding, subtracting, multiplying, and dividing fifth grade level problem
- Level 2 – Green Cards: Adding, subtracting, multiplying, and dividing fifth grade level problems that are harder than Level 1
- Level 3 – Red Cards: Adding and subtracting six digit numbers with carrying and borrowing, and multiplying and dividing multi-digit problems

$$3 + 4$$

$$7 + 3$$

4
+ 3
———
7

3
+ 7
———
10

5
+ 5

7
+ 4

5
+ 5
―――
10

4
+ 7
―――
11

$$9 + 5$$

$$9 + 5$$

5
+ 9

14

5
+ 6

11

$$9 \atop +\ 5$$

$$9 \atop +\ 2$$

$$\begin{array}{r} 6 \\ +\ 5 \\ \hline 11 \end{array}$$

$$\begin{array}{r} 6 \\ +\ 2 \\ \hline 8 \end{array}$$

5
+ 7

8
+ 6

$$\begin{array}{r} 7 \\ +\ 5 \\ \hline 12 \end{array}$$

$$\begin{array}{r} 6 \\ +\ 8 \\ \hline 14 \end{array}$$

8
+
7

6
+
7

$$\begin{array}{r} 7 \\ +\ 8 \\ \hline 15 \end{array}$$

$$\begin{array}{r} 7 \\ +\ 6 \\ \hline 13 \end{array}$$

$$\begin{array}{r} 8 \\ +\ 3 \\ \hline 11 \end{array}$$

$$\begin{array}{r} 7 \\ +\ 9 \\ \hline 16 \end{array}$$

$$\begin{array}{r} 3 \\ +\ 8 \\ \hline \end{array}$$

$$\begin{array}{r} 9 \\ +\ 7 \\ \hline \end{array}$$

8 + 8

8 + 9

$$\begin{array}{r} 8 \\ + 8 \\ \hline 16 \end{array}$$

$$\begin{array}{r} 8 \\ + 6 \\ \hline 14 \end{array}$$

$$9 + 6$$

$$9 + 3$$

9
+ 6

15

9
+ 3

12

$$8 + 9$$

$$7 + 9$$

$$\begin{array}{r} 9 \\ +\ 8 \\ \hline 17 \end{array}$$

$$\begin{array}{r} 9 \\ +\ 7 \\ \hline 16 \end{array}$$

$$18 + 2$$

$$9 + 9$$

$$\begin{array}{r} 18 \\ + 2 \\ \hline 20 \end{array}$$

$$\begin{array}{r} 9 \\ + 9 \\ \hline 18 \end{array}$$

11 ÷ 4 = 1

11 ÷ 3 = 1

11
4
- —
7

11
3
- —
8

11
- 7
———

11
- 9
———

11
− 7
——
4

11
− 6
——
5

12 − 5

12 − 4

12
- 5
———
7

Project MIND
Second - Yellow

12
- 4
———
8

Project MIND
Second - Yellow

1   13 | 4

1   12 | 9

13
- 4
___
9

12
- 9
___
3

13
- 8

13
- 6

13
- 8
5

13
- 6
7

$$14 - 6 = \underline{}$$

$$13 - 6 = \underline{}$$

14
- 6
8

13
- 9
4

$$14 - 8$$

$$14 - 7$$

14
− 8
_____
6

14
− 7
_____
7

15 | 5
1

14 | 9
1

$$\begin{array}{r} 15 \\ -\ 5 \\ \hline 10 \end{array}$$

$$\begin{array}{r} 14 \\ -\ 9 \\ \hline 5 \end{array}$$

1
- 7
15

1
- 6
15

15
- 7
8

15
- 6
9

16

2

−

15

8

−

```
  16
-  2
────
  14
```

```
  15
-  8
────
   7
```

$$\frac{1}{8} \quad 16$$

$$\frac{1}{7} \quad 16$$

16
- 8
8

16
- 7
9

$$\begin{array}{r} 17 \\ -\ 7 \\ \hline \end{array} \qquad \begin{array}{r} 16 \\ -\ 9 \\ \hline \end{array}$$

17
7
——
10

Project MIND
Second - Yellow

16
9
——
7

Project MIND
Second - Yellow

18
- 9

17
- 8

18.

9
9

17.

8
9

$$28 + 5$$

$$29 + 4$$

$$\begin{array}{r} 5 \\ +\ 28 \\ \hline 33 \end{array}$$

$$\begin{array}{r} 4 \\ +\ 29 \\ \hline \boxed{33} \end{array}$$

15
+ 18

15
+ 9

$$\begin{array}{r} 15 \\ +\ 18 \\ \hline 33 \end{array}$$

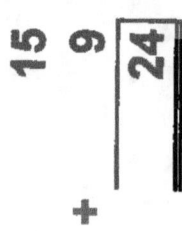

$$\begin{array}{r} 15 \\ +\ 9 \\ \hline 24 \end{array}$$

$$16 + 27$$

$$16 + 8$$

```
   16
+  27
-----
   43
=====
```

```
   16
+   8
-----
   24
```

$$26 + 18$$

$$25 + 65$$

```
    26
+   18
   ————
    44
```

```
    25
+   65
   ————
    90
```

$$\begin{array}{r} 28 \\ + 26 \\ \hline \end{array}$$

$$\begin{array}{r} 26 \\ + 64 \\ \hline \end{array}$$

28
26
+ ———
54

26
64
+ 90

$$\begin{array}{r} 29 \\ +\ 13 \\ \hline \end{array}$$

$$\begin{array}{r} 28 \\ +\ 47 \\ \hline \end{array}$$

```
  29
+ 13
━━━━
  42
```

```
  28
+ 47
━━━━
  75
```

35
+ 8

29
+ 57

$$35 + 8 = 43$$

$$29 + 57 = 86$$

$$\begin{array}{r} 38 \\ +\ 25 \\ \hline \end{array}$$

$$\begin{array}{r} 36 \\ +\ \ 7 \\ \hline \end{array}$$

```
    38
  + 25
  ————
    63
```

```
    36
  +  7
  ————
    43
```

```
  39        39
+ 21      + 11
```

```
  39
+ 21
————
  60
```

```
  39
+ 11
————
  50
```

```
  45
+  8
————
```

```
  44
+ 26
————
```

$$\begin{array}{r} 45 \\ +8 \\ \hline 53 \end{array}$$

$$\begin{array}{r} 44 \\ +26 \\ \hline 70 \end{array}$$

$$\begin{array}{r} 46 \\ +7 \\ \hline \end{array}$$

$$\begin{array}{r} 45 \\ +25 \\ \hline \end{array}$$

$$\begin{array}{r} 45 \\ +\ 25 \\ \hline \boxed{70} \end{array}$$

Project MIND
Second - Green

$$\begin{array}{r} 46 \\ +\ 7 \\ \hline 53 \end{array}$$

Project MIND
Second - Green

$$\begin{array}{r} 48 \\ +\ 14 \\ \hline \end{array}$$

$$\begin{array}{r} 46 \\ +\ 35 \\ \hline \end{array}$$

```
  48
+ 14
―――
  62
```

```
  46
+ 35
―――
  81
```

$$\begin{array}{r} 49 \\ +\ 18 \\ \hline \end{array}$$

$$\begin{array}{r} 48 \\ +\ 26 \\ \hline \end{array}$$

```
   49
+  18
─────
   67
═════
```

Project MIND
Second - Green

```
   48
+  26
─────
   74
```

Project MIND
Second - Green

$$\begin{array}{r} 58 \\ +\ 19 \\ \hline \end{array}$$

$$\begin{array}{r} 47 \\ +\ 26 \\ \hline \end{array}$$

```
   58
+  19
-----
   77
```

```
   47
+  26
-----
   73
```

```
  64
+ 19
─────
```

```
  63
+ 29
─────
```

$$
\begin{array}{r}
64 \\
+ \ 19 \\
\hline
83
\end{array}
$$

$$
\begin{array}{r}
63 \\
+ \ 29 \\
\hline
92
\end{array}
$$

$$\begin{array}{r} 67 \\ +\ 24 \\ \hline \end{array}$$

$$\begin{array}{r} 65 \\ +\ 8 \\ \hline \end{array}$$

```
  67
+ 24
─────
  91
```

```
  65
+  8
─────
 73
```

$$75 + 9$$

$$68 + 13$$

```
  75
+  9
─────
  84
═════
```

```
  68
+ 13
─────
  81
═════
```

32
- 15

27
- 8

```
  32
- 15
─────
  17
═════
```

```
  27
-  8
─────
 │19│
═════
```

34 - 7

33 - 15

```
  34
-  7
  27
```

```
  33
- 15
  18
```

35 - 15

35 - 7

```
   35
-  15
   20
```

```
   35
-   7
   28
```

43
- 19

39
- 18

43
- 19
———
24

39
- 18
———
21

$$\begin{array}{r} 46 \\ -\ 27 \\ \hline \end{array}$$

$$\begin{array}{r} 45 \\ -\ \ 6 \\ \hline \end{array}$$

46
27

19

45
- 6

39

51 - 14

46 - 28

```
  51
- 14
-----
  37
```

```
  46
- 28
-----
  18
```

56
- 19

53
- 18

$$\begin{array}{r} 56 \\ -\ 19 \\ \hline 37 \end{array}$$

$$\begin{array}{r} 53 \\ -\ 18 \\ \hline \boxed{35} \end{array}$$

61
- 4
_____

57
- 29
_____

61
- 4
57

Project MIND
Second - Green

57
- 29
28

Project MIND
Second - Green

$$\begin{array}{r} 62 \\ -\ 59 \\ \hline \end{array}$$

$$\begin{array}{r} 61 \\ -\ 13 \\ \hline \end{array}$$

62
59
3

61
13
48

66
- 47
_____

64
- 27
_____

```
  66
-  47
-----
  19
```

```
  64
-  27
-----
  37
```

$$71 - 28$$

$$71 - 5$$

71
- 28
———
43

71
- 5
———
66

$$
\begin{array}{r}
80 \\
-\ 26 \\
\hline
\end{array}
\qquad
\begin{array}{r}
71 \\
-\ 39 \\
\hline
\end{array}
$$

$$\begin{array}{r} 80 \\ -\ 26 \\ \hline 54 \end{array}$$

$$\begin{array}{r} 71 \\ -\ 39 \\ \hline \boxed{32} \end{array}$$

81 | 9
___

80 | 40
___

```
  81
-  6
────
  75
```

```
  80
- 40
────
  40
```

$$1\,\overline{)\,89}$$

98

$$1\,\overline{)\,36}$$

90

98
− 89
———
9

90
− 36
———
54

$$\begin{array}{r} 99 \\ -\ 19 \\ \hline \end{array}$$

$$\begin{array}{r} 99 \\ -\ 68 \\ \hline \end{array}$$

99
- 19
———
80

99
- 89
———
10

$$150 + 239$$

$$150 + 129$$

$$\begin{array}{r} 150 \\ +\ 129 \\ \hline 279 \end{array}$$

Project MIND
Second - Red

$$\begin{array}{r} 150 \\ +\ 239 \\ \hline 389 \end{array}$$

Project MIND
Second - Red

$$\begin{array}{r} 212 \\ +\ 173 \\ \hline \end{array}$$

$$\begin{array}{r} 202 \\ +\ 183 \\ \hline \end{array}$$

212
+ 173
───
385

202
+ 183
───
385

```
  226
+ 527
```

```
  217
+ 543
```

$$\begin{array}{r} 226 \\ + \ 527 \\ \hline 753 \end{array}$$

$$\begin{array}{r} 217 \\ + \ 543 \\ \hline 760 \end{array}$$

$$327 + 246$$

$$326 + 147$$

```
    327
  + 246
  ─────
    573
  ═════
```

```
    326
  + 147
  ─────
    473
  ═════
```

$$\begin{array}{r} 382 \\ +\ 408 \\ \hline \end{array}$$

$$\begin{array}{r} 372 \\ +\ 418 \\ \hline \end{array}$$

```
    382
 +  408
    ___
    790
    ===
```

```
    372
 +  418
    ___
    790
    ===
```

$$\begin{array}{r} 417 \\ +\ 265 \\ \hline \end{array}$$

$$\begin{array}{r} 417 \\ +\ 253 \\ \hline \end{array}$$

417
+ 265
—————
682

417
+ 253
—————
670

$$563 + 29$$

$$553 + 28$$

$$\begin{array}{r} 563 \\ +29 \\ \hline 592 \end{array}$$

$$\begin{array}{r} 553 \\ +28 \\ \hline 581 \end{array}$$

$$
\begin{array}{r}
615 \\
+\ 28 \\
\hline
\end{array}
\qquad
\begin{array}{r}
612 \\
+\ 29 \\
\hline
\end{array}
$$

$$\begin{array}{r} 615 \\ +28 \\ \hline 643 \end{array}$$

$$\begin{array}{r} 612 \\ +29 \\ \hline 641 \end{array}$$

$$622 + 39$$

$$621 + 154$$

```
  622
+  39
-----
  661
```

```
  621
+ 154
-----
  775
```

$$632 + 254$$

$$624 + 37$$

```
  632
+ 254
─────
  886
═════
```

```
  624
+  37
─────
  661
═════
```

```
  715
+  29
_____
```

```
  705
+  39
_____
```

```
  715
+  29
─────
  744
```

```
  705
+  39
─────
  744
```

$$876 + 115$$

$$9998 + 107$$

```
  876
+ 115
─────
  991
```

```
  866
+ 107
─────
  973
```

235 - 17

224 - 17

235
- 17
218

224
- 17
207

$$342 - 41$$

$$332 - 32$$

342
- 41
———
301

332
- 32
———
300

$$462 - 128$$

$$447 - 326$$

462
- 128
334

447
- 326
121

466 − 38

463 − 128

$$466 - 38 = 428$$

$$463 - 128 = 335$$

476
- 38

467
- 325

$$\begin{array}{r} 476 \\ -38 \\ \hline 438 \end{array}$$

Project MIND
Second - Red

$$\begin{array}{r} 467 \\ -325 \\ \hline \boxed{142} \end{array}$$

Project MIND
Second - Red

$$
\begin{array}{r}
581 \\
- 233 \\
\hline
\end{array}
$$

$$
\begin{array}{r}
580 \\
- 234 \\
\hline
\end{array}
$$

581
- 233
348

580
- 234
346

```
  635
-
  318
_____
```

```
  625
-
  317
_____
```

```
   635
 - 318
 ‾‾‾‾‾
   317
```

```
   625
 - 317
 ‾‾‾‾‾
   308
```

```
      672          662
    - 371        - 361
```

$$\begin{array}{r} 672 \\ -\ 371 \\ \hline 301 \end{array}$$

$$\begin{array}{r} 662 \\ -\ 361 \\ \hline 301 \end{array}$$

$$767 - 53$$

$$763 - 253$$

```
  767
-  53
 ────
  714
```

```
  763
- 253
 ────
  510
```

787
− 71

767
− 245

```
    787
  -  71
    716
```

```
    767
  - 245
    522
```

```
  818        797
- 502      - 231
-----      -----
```

818
502
316
-

797
231
566
-

```
    858
  - 415
```

```
    840
  - 230
```

```
  858
-  415
  443
```

```
  840
-  230
  610
```

```
    873
-   263
─────────

    866
-   414
─────────
```

```
  873
- 263
─────
  610
```

```
  866
- 414
─────
  452
```

$$928 - 605$$

$$875 - 404$$

928
605
———
323

875
404
———
471

$$966 - 450$$

$$956 - 133$$

960
- 450
510

956
- 133
823

```
  976
- 352
_____
```

```
  9668
-  405
_____
```

976
352
624

968
405
563

$$997 - 355 = $$

$$987 - 354 = $$

```
  997
- 355
─────
  642
═════
```

```
  987
- 354
─────
  633
```

www.ingramcontent.com/pod-product-compliance
Lightning Source LLC
Chambersburg PA
CBHW080247180526
45167CB00006B/2447